The Saguaro Forest

The Saguaro Forest

Text by PETER WILD · Photographs by HAL COSS

NORTHLAND PRESS · FLAGSTAFF, ARIZONA

The legend beginning on page 36 appears courtesy the
Arizona Quarterly. "Papago Legend of the Sahuaro," by Susie
Ignacio Enos, *Arizona Quarterly* 1 (1945):64-69.

The legend beginning on page 43 was first published as
"The Giant Cactus," in *Long Ago Told: Legends of the Papago
Indians,* by Harold Bell Wright (New York: D. Appleton
& Co., 1929), 109-22.

The photographs on pages 28 and 29 were taken by Helga
Teiwes and are used with permission.

Contents

Introduction 1

Origins: From Forest to Forest 5

Origins: Wine, Rain, and the Staff of Life . .22

"Looking for My Mother" 35

The Future: The Great Saguaro Debate 49

Further Reading 63

# Introduction

Americans rarely make much to-do about state flowers. That Alaska's is the forget-me-not seems cute in a touristy sort of way. That Missouri's is the hawthorn, or that the mountain laurel serves as Connecticut's emblem (the same plant does double duty for nearby Pennsylvania), may strike us as less than gripping. What, in fits of local patriotism, state legislators bless with official sanction to emblazon maps and tourist literature is soon forgotten, if thought much about in the first place.

Beyond that, one might make a case that Michigan's apple or Delaware's peach blossoms not only memorably beautify their states in the spring but symbolize major contributions to their economies. Yet strangely enough, perhaps the best case for an appropriate state flower concerns one that few people see. It blooms at night, often miles from human habitation, then withers the following afternoon. It appears atop rather bizarre plants, "enormous columns of vegetable matter," as one Englishman described them in the last century. They dot the southern deserts of Arizona and northwestern Mexico, mature specimens rising from thirty-five to fifty feet and covered, not with leaves, but with thousands of spines shining in the sunlight.

No wonder these green masses looking like huge candelabra have struck various travelers over the years as grotesque, unique, and curious. And so they may appear today to desert visitors who first see them from their car windows. Marching up and

1

Saguaro blossom

down bouldery mountains and out into valleys, they dominate the bare landscape — weird vegetation from another planet. Yet there is nothing strange about saguaro cacti. They only seem that way because of the prejudices we bring to them, our predilections for the "ordinary" trees and shrubs and flowers that make up the "natural" landscape that we carry around in our heads. In truth, no matter what adjectives we apply to it, the saguaro, like other plants, is a product of its environment and thus supremely adapted to it. The saguaro appears to be different only because its dry, hot, rocky home in the Southwest differs remarkably from most areas in the United States.

Although the saguaro serves little economic importance today, and provides mostly curious, rather than aesthetic, allure for camera–toting travelers, its blossom makes an exemplary state emblem. For, as we shall see, this cactus, arresting for its size and shape, not only holds out a key to the whole complex but delicate web of desert ecology, it can open windows as well on the region's history and Indian lore. A scientific look at the enormous plant, fascinating in itself, provides a measure of the region's future. With this in mind, it is not too much to say that to learn about the saguaro is to learn a good deal about the desert in which it lives.

Origins: From Forest to Forest

In their daily lives, Indians showed an awareness of environmental nuances that few people today can match. Curiously, however, their myths transform everyday diversity into grand stereotypes. Thus, though no doubt keen-eyed Papagos could recognize individual coyotes who made regular rounds to village gardens, in legend coyotes become a single archetype: Coyote, a demigod who stands as the representative of all "coyotedom." A similar observation holds for other characters in the mythical menagerie. Skunk, Mockingbird, and Badger speak like men tinged with magic as they walk on,

Prickly pear cactus in the foreground, with saguaros in the background.

fly over, or burrow into the earth. The perceived timelessness of nature, the rocky hills always recognizably there, the stars that repeat their patterns season after season in the same paths, fostered the attitude. Thus in their myths, the Papagos look on the giant cactus as godlike, unchanging.

In a similar way, newcomers to southern Arizona and northern Mexico may share the static view of the largest cactus within United States borders. In places, the traveler can drive for mile after mile thick with these huge plants. They rise beyond the windshield, rank after endless rank of spiny columns so dense that they lose their individuality. They so dominate the landscape that the word godlike naturally comes to mind.

5

In such a way, the species becomes elevated in the popular imagination to an ill-defined but prevalent heroics. The business portion of the Tucson telephone directory contains some twenty entries under "Saguaro," the Phoenix directory nearly seventy — not to mention listings under variant spellings. The saguaro is thus recruited to serve as the emblem for everything from art galleries to pest-control companies.

Yet visitors who take time out to park their cars and stroll around in this unusual forest until they become somewhat attuned to the surroundings will begin to notice surprising individual differences. They hike up a hill thick with healthy saguaros, only to reach the top and peer beyond over a long slope nearly bare of the giants. They think back on their high school botany and scratch their heads. It is difficult to imagine that the rainfall, sunlight, soil, and other factors controlling growth change that radically with the distance it takes to crest a ridge. Furthermore, in some parts of the forest visitors note stands of cacti all roughly the same height. In patches but a few miles away they find saguaros ranging from knee-high youngsters to towering grandfathers.

Inspection of individual plants only compounds the puzzle. One saguaro stands abruptly narrowed in its middle, as if a great hand had pinched a great green straw. Another, though looking normal enough otherwise, is dotted with holes and black, unsightly scars. One saguaro branches out with arms twisted as from torture, while another exhibits a peculiar formation at its top, a bizarre, green crown. Then the desert wanderer stumbles on a giant seemingly exploded from within, amputated arms strewn wildly over the ground, and the remaining stump dripping a putrid liquid.

The desert, popularizers keep telling the public, is a place full of surprises. In books and films we see that it is a land where minute shrimp blow on the wind, where seeds may lie dormant for years awaiting the precise combination of favorable conditions to sprout, where on rare occasions the cross-country walker may chance upon pelicans stumbling about the desert scrub just as bewildered as the people who see them, and where one might come upon the contradiction of a palm tree far back in an arid canyon. Perhaps most astonishingly in this place of brimstone, some mammals, such as the kangaroo rat, eschew drinking water throughout their lives.

The avid desert barker could compile a book-length list of believe-it-or-nots. But to appreciate the desert as a whole, we should beware the Disneyland syndrome. All the items above do occur, though usually they go unnoticed by all but the most patient and observing humans. In the larger context, what appear to be incongruities should not be seen as separate, freakish events of nature, but as parts of the ordinary energy flow of everyday life.

For if the hardwood stretches of New York State or the evergreen forests of the Pacific Northwest seem in some way humdrum by comparison to the seekers of nature's thrills, such people should also realize that desert life is in no wise otherworldly. It is diverse because its environment is diverse, demanding extreme adjustments to extreme conditions. This holds true no more for the saguaro than for any other plant or animal. For the giant cactus, however, the drama is more obvious to humans because of the plant's size and more biologically significant to the whole because this great column stands dominant in its range, at the center of the region's ecological web.

Having said that, it comes as some surprise that these Gargantuas of the plant world are relative upstarts in the scale of evolution. Saguaros, of course, belong to the cactus family, a large New World category boasting hundreds of separate species. Ranging from the picayune to the colossal, they dot both North and South America from Patagonia to the Northwest Territories of Canada. Despite the exuberant varieties in shape, color, and size, cacti share common family traits. Nearly all of these perennials have fleshy stems exhibiting scales or spines instead of leaves. On the score of identification, however, a passing note of caution is in order. All that bristles is not a cactus.

Ascribing to themselves a manly ruggedness to match a rugged land, old-timers bragged about Arizona as the devil's cactus-studded domain. "Hell's Half Acre," a poem credited to a Tucson saloon keeper of a century ago, has entered the folklore: "The devil was given permission one day / To select a land for his own special sway. . . ." As the poem develops, Satan chooses bleak Arizona as the place dearest to his heart. Then he decides to redecorate the sandy waste, fuming and swearing up a fit as he furnishes his new realm to his liking with rattlesnakes, tarantulas, and other manner of creeping, stinging vermin. Cacti, however, are his special delight:

> He saw there were still improvements to make
> For he felt his own reputation at stake.
> An idea struck him, he swore by his horns
> That he would make a complete vegetation
>   of thorns.
> So he studded the land with the prickly pear
> And scattered the cacti everywhere,
> The Spanish dagger, pointed and tall,
> And at last the cholla to outstick them all.

"Everything has either horns or thorns," was also once commonly boasted in the Southwest. That does not mean, however, that everything that pricks is attached to a cactus, any more than everything with horns is a cow. We should remind ourselves that Arizona's diversity includes impressive swaths of pine forests in the central and northern parts of the state. True, the southern region sports a panoply of stiletto-armed plants, as the unwary hiker will

Storm approaching over a solitary saguaro

soon discover. But the saloon keeper is a bit misleading not only in his geography but in his botany. The mentioned Spanish dagger, or Spanish bayonet as it is more commonly called, though cactuslike, is a yucca often classified as belonging to the lily family. All of which is to point out that many unrelated species have struggled through their generations with the problems of water conservation and protection from predators. Sometimes, they arrived at similar solutions. In any case, as the largest cactus on this side of the border, the mature saguaro presents few problems in identification, even to novices.

Yet the saguaro's family origins, that is, the ancestor of the entire cactus clan, still puzzles specialists. Unfortunately, cacti show up sparsely in the fossil record; here, the oldest evidence goes back only about 40,000 years, a mere blink at the parade of the eons. Scientists do note that in spite of their extensive range, cacti flourish best in warm climates, especially in the desert borderlands adjacent to the tropics. One theory has it that this fairly recent family began millions of years ago in the hot and humid West Indies. There, a long-term drying trend started delicate violets and begonias on their epochal march of adjustment. Whatever the evolutionary details, perhaps forever lost to science, two certainties in this development stand out. The saguaro continues to evolve, a factor of prime consideration when we turn to its future in the Southwest, and the saguaro's foremost adaptation relates to water scarcity.

It is actually a series of adaptations that are reflected in the saguaro's structure. A dull, waxy sheen, or cuticle, covers the entire plant. This helps prevent water loss. The broad foliage of humid climates may give us aesthetic pleasure, a sense of lushness and well-being on a summer's day, but such structures surrender moisture rapidly. In a hot and arid region, they can be unaffordable luxuries. Most perennial desert plants do have leaves, but evolution has modified them radically, reducing their size and giving them waxy patinas in order to slow transpiration, or water loss. Furthermore, plants such as the ocotillo shed their leaves during dry periods. To dispense with leaves entirely, as most cacti have done, seems an even better expedient.

Still, serving as the main food-manufacturing area of most plants, leaves fulfill essential functions. Through an unusual ploy, the leafless saguaro gains the best of two worlds. Over its long development the cactus has transferred chlorophyll, the green, food-making matter in leaves, to its trunk. Lying under the protective, waxy coating, the chlorophyll goes about its job without sapping the plant's water reserves.

That job involves a good measure of complexity and keen adjustment to arid conditions, for the plant needs to draw on the dry, outside air to make its food—on the very air from which the waxy cuticle seals it off. To accomplish this with a minimum of water loss, the stomata, or pores, of the trunk stay closed during the day. They open during the lower temperatures of night, permitting carbon dioxide from the surrounding air to enter. At this point the saguaro would seem to face another problem: photosynthesis can occur only with the help of light.

With another deft trick, the saguaro converts the carbon dioxide into substances held over for use the next day. This feature, shared by other succulent plants, goes by the term crassulacean acid metabolism (CAM), a name that slips poetically over the tongue despite its length.

But the plot thickens, for there is still more to this tale of adjustment. The saguaro stores the carbon dioxide by turning it into an acid, a process favored by the low nighttime temperatures. The lower the temperature, the more acid produced. By the morning after a cool night, the saguaro is chockfull of acid waiting to be converted into food with the rising of the sun. Because of this, the saguaro, together with most other cacti, thrives in regions with high day-night temperature differences. The condition holds eminently true for the Sonoran Desert. Here summer days are hot, but evenings cool off as the dry, cloudless nights allow the day's heat to rise into the atmosphere.

At first thought we might dismiss the thousands of inch-long spines studding a saguaro as so much primitive armor. Needle sharp as they are, they serve this function well. In a "devil's domain," where nearly every creature prowls about in a competitive search for moisture, few large animals nibble at the succulent but bristling saguaro. Yet the spines do more than yeoman's service as the forward guard of the plant's water reservoir. In a shadow-scarce land, they shade the tall, exposed plant, and their ranks of thousands lessen evaporation by breaking the force of the moisture-sucking desert winds. The stabbers may provide yet another and more subtle advantage. One theory has it that the spines discharge electricity from their points and by releasing ions help lift the water within a plant so tall that capillary action alone cannot do the job.

Bringing to mind the fluted columns of ancient Greek architecture, vertical ridges run top to bottom on saguaros. These do more than bear the multiple-functioning spines. The saguaro is dynamic, and in a supreme adjustment to water scarcity, its accordion pleats allow the entire above-ground plant to expand and contract. Hence, it swells as it absorbs water during the rainy seasons, then shrinks as it draws on its own stored sap to get through the dry seasons of fall and spring. Some summers, as the Papago farmers know too well, the thunderheads let down only a fraction of their promise. But the saguaro, containing up to ninety percent water, holds enough to last the cactus through dry years. This feature has led to speculation that the giants tip the scales at from six to eight tons—even more by some guesses. People fascinated by superlatives will be crestfallen at the actual figures. In 1969 park service employees at

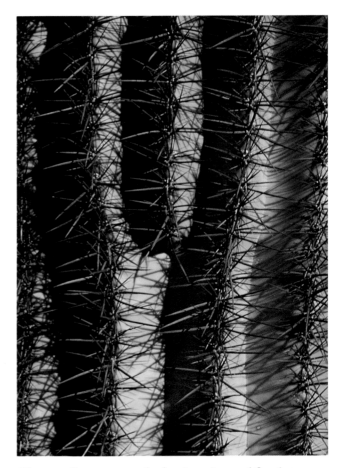

Close-up of a saguaro trunk, showing spines and fluted structure. The "accordion" trunk expands when the plant is taking up moisture and contracts during the dry season.

Saguaro National Monument near Tucson removed two mature saguaros threatening a house at the Madrona Ranger Station. Hoping that "the destruction of these plants would not be a total loss," officials invited a professor and his students from the University of Arizona to weigh the downed specimens and settle the question once and for all. The finding: one tallied 2,500 pounds, the other 1,600 pounds, far cries from previous estimates.

The numbers are impressive nevertheless, and they have much to say about the saguaro's internal workings as an elongated water-storage device. In cross-section, the cactus displays a fairly simple structure. The mass of tissue consists of water-absorbing pulp. Within this rises a circle of from twelve to twenty slender rods. These ribs not only add rigidity to the soft but soaring mass, they assist the vascular bundles in carrying nutrients throughout the system. Often, long after the fleshy part of the plant has died and withered to dust, they remain standing in their original clustered shape, the saguaro's bleached skeleton. These long "bones" are ribs prized by the Indians as building materials.

Tales in folklore tout the saguaro as nature's version of the emergency water tank. Subsequently, most Americans carry around in their heads the picture of the old sourdough crawling across the sands. He spies a cactus on the horizon and speedily grovels his way toward it. Arriving, he whips out his Bowie knife, cuts a hole in the trunk, and drinks deeply from the exuding liquid.

It's an appealing fable. True, the saguaro is a column of water held up by supporting tissue, but in this regard its design serves itself better than it does man. No gushing spigot can be cut into its side, for, again to preserve easy water loss, the sap is thick, mucilaginous. Supposedly, the procedure is to cut the top from any barrel-shaped cactus of convenient height and pound the exposed mass into a puddle. One could also cut a hole in a tall cactus and jab away with an inserted stick until liquid flows, or slice off a few chunks for chewing. The ironic result: a mean-tasting, unpalatable, and illness-producing sap. Further, the plant destruction is strictly illegal.

In any case, the abused cactus will not necessarily die, unless its top is lopped off. Admirably ready for the slings and arrows of its environment, the saguaro exudes a liquid. This hardens into a callus to seal off wounds. Thus the dark scars on saguaros along many roadsides are the plants' versions of permanent scabs coming to the aid after injuries. So hardy is the saguaro in this respect that even a hole excavated deep into the central tissue by a bird for a nesting spot secure from predators will not kill a saguaro, though it might weaken the column for wind and freeze damage. In fact, the giant cactus develops a protective callus so hard that it remains long after the spiny tree has fallen. Picked out of the plant's debris, these former linings of cavities pecked out by birds provide the globular "saguaro boots" treasured by human curiosity seekers. Despite their name they are not used as shoes, although in times past, Papago children picked them up and carried them home for doll houses.

In a different response to damage, saguaros develop bizarre, convoluted crowns at their tops. Though the matter is by no means settled, botanists think that these result from mutations, a fungus, or from mechanical injury, as when the saguaro's growing tip wears against the overhanging limb of a tree. Rare but eye-catching, these crested, or cristate, forms are worth looking for during a stroll through the cactus forest.

Like that of other plants, the life of the saguaro goes on underground as well as above. Roots gather nutrients and anchor the growth above. Basically, vegetation responds to these needs with one of two patterns. The mesquite tree common to southwestern deserts, for instance, sends a primary root plunging deep into the earth. The long root taps subterranean moisture lying far below the surface. It also gives the tree a stubborn grasp on terra firma, as people laboring to remove mesquite stumps can attest.

Along with many other arid-land plants, the saguaro exhibits a different system. Instead of diving deep, its roots spread out to suck as much moisture as possible falling on the earth's surface. In the case

Crowned, or cristate, saguaro — a puzzling shape sometimes found in the saguaro forest.

of the saguaro, this network radiates from the base for a distance roughly equal to the cactus's height.

The saguaro's strategem for skimming water from the surface meets with remarkable success. It is not true, to deflate yet another yarn that some old-timers stoutly propound, that saguaro roots are so greedy for water that they can lap up even the scanti-est of rainfalls. A shower of less than 0.20 of an inch probably will evaporate back into the dry atmos-phere before most vegetation, saguaros included, can get to it. Nevertheless keen for moisture, the saguaro's underground network grows temporary rain roots after an adequate storm. These probe out to gather the dampness soaking in and transport it to the permanent roots. Then they wither away.

The intricate but shallow circle of roots comes with one major drawback. Ironically, and to their downfall, saguaros can get too much water. As with any desert plant, overabundance can kill the cactus. However, in a place such as Tucson, Arizona, which receives about eleven inches of rain a year—and sometimes less—this occurs more often from exuber-ant hose-work by suburbanites than from natural overabundance. The saguaro, as it happens, prefers coarse, well-drained soil, yet it consists of a huge, cigar-shaped water tank balanced on end and held upright only by a shallow lacework of underground guy wires. If growing in sandy soil, soil that becomes thoroughly soaked, the monster's delicate anchors

can fail, especially in a high wind after a good rain.

The cactus is irreplaceable, for common wisdom rightly tells us that cacti grow slowly. Just how slowly is not a matter easily determined. It would be convenient to have a figure of, say, one inch, or one foot, or whatever, a year to tuck away in our minds for easy reference. Even if we had such a figure, it would be an oversimplification of a complex process of nature's human-defying clocks. The disheveled fact is that the growth rates of individual saguaros vary, and the rate of any one plant depends on a host of environmental aspects working in combination: availability of moisture, exposure to sunlight, soil type, temperature fluctuations over its lifetime, and other influences. Furthermore, the saguaro grows at different rates in different periods of its life, thus ruining hopes of establishing a rule-of-thumb figure.

The saguaro has yet another tic in the matter of age, for the giant cactus changes shape in the course of its development. Beginning as a tiny, green spike, it grows into an egglike or globose stem, stretches out into a club, then stretches farther to assume a bowling-pin form narrow at the top. Lastly, it settles into what we usually think of as the typical saguaro outline, beloved by artists and photographers, of the branched adult.

The progressive changes taking place in a saguaro's life are not so much related to age as they are to size. In other words, saguaros of uniform

14

Top left: Saguaro seedling only a few days old. Top right: One-year-old saguaro. Bottom left: Three-year-old saguaro. Bottom right: Club-shaped saguaro about six years old. Note the camera lens cap used for size comparison. Above: Club-shaped young saguaros under a palo verde "nurse tree." A senita cactus grows behind the saguaro cluster.

height can differ radically in age. This makes sense. If, for example, a saguaro lingers stunted because of highly unfavorable soil and moisture conditions, it may never bloom, while a much younger specimen, better blessed, is putting out orgies of blossoms. All this takes place in a life normally spanning 150 to 175 years.

Despite the convolutions of nature's ways, the situation does not leave us entirely shipwrecked in uncertainty. To generalize—and to emphasize that experts' figures vary—after one year of growth, a seedling may be an all but unrecognizable tuft barely one-quarter of an inch above the ground. Once it becomes established, however, the plant enters a period of rapid growth. By its twenty-first year, it may be all of a yard high; by its thirtieth year, two yards. Sometime between then and its fortieth season, when the cactus rivals or surpasses the height of a man, it begins its reproductive life. A few tentative blossoms appear at its top, preparation for the increasing numbers of blossoms of maturity. At about fifteen feet and roughly seventy-five years of age, the individual develops arms. Eventually, both the top and arm tips bear hundreds of creamy white flowers. Yet the saguaro sacrifices something for the gain. With energy now going into reproductive effusion, stem growth slows in mature adults.

Be that as it may, nature does not measure success by height but by the ability to produce offspring, which in turn succeeds in fulfilling the pattern ad infinitum. According to this urge, from April to June, depending on when the warming trend of spring starts, individual saguaros hold out their trumpet-shaped flowers. Thus they begin, as we shall see, to run the gauntlet of reproductive succession. The petals open after the sun sets and close the following afternoon. Over the years, there has been some debate over what pollinates these night bloomers. The wind is not strong enough to carry their heavy pollen. Though honey bees cloud around the cacti, they were introduced to the Southwest only a little over a hundred years ago, long after the successful pollination of existing saguaros.

Nature takes the scattergun approach, pollinating the cactus through a variety of creatures. A number of insects help. Doves perch delicately on saguaros to sip nectar, in the process spreading pollen borne on their beaks and feathers from plant to plant. But these and other birds contributing to the process fly about during the day. Pollination of saguaros depends mainly on nocturnal creatures such as the long-nosed bat. This migrant, who arrives in the saguaro's homeland around the blooming season, has the long tongue typical of a species with a nectar-specific diet. Experiments offer further convincing evidence. A larger proportion of flowers set fruits after visits from long-nosed bats than from those of birds or insects.

The ripe, plum-sized fruits start ripening in June, then begin to fall. They weigh just under two ounces

The red fruits of the saguaro are often mistaken for blossoms.

each but contain 2,000 or more tiny seeds. The fruits' juicy, bright red pulps spark off a commotion across the saguaro forest. All life seems to pounce on them as the narrowing race for survival begins. Ants, doves, pocket mice, kangaroo rats, thrashers, ground squirrels, coyotes, javelina, and quail flock in to pick up the saguaro's succulent manna. To illustrate the feasting frenzy, scientists find that at this time of year coyote scat contains about ninety-five percent saguaro seeds, and javelina scat about eighty-five percent.

Within a few hours after the ripe fruits fall, there may be a seed shadow at the base of the parent cactus, an area littered with husks but no seeds or pulp. Yet some of the genetic freight from the fruit falls into rocky crevices, the wind blows some out of reach, and a storm may wash seeds away from the banquet to an isolated spot. Seeds also pass unharmed through some animals' digestive tracts, to be dropped far from where they originally fell to the ground. In sum, nature again uses the scattergun approach in the process of producing another generation, for of a thousand seeds eventually coming to rest at a favorable germination site, less than one percent will survive to become established plants.

Of these, fewer still will survive nature's lottery. Not only does the gauntlet of hungry animals take a heavy toll, the seeds themselves require a specific set of combined conditions. Germination coincides with the monsoon of July and August. But the seeds

need not only moisture, the right soil, and warm temperatures to sprout. They do best with at least two rain storms within a period of a few days—and the monsoon is irregular. Young saguaros also do best when growing under trees or shrubs, and random dispersal favors those seeds that land in such protected areas. Shade moderates the microenvironment throughout the plants' early years, decreasing the searing heat, retaining soil moisture, screening from frost, and offering protection from trampling animals. In one spectacular example, eighty-three young saguaros clustered in and around the shelter of one such "nurse tree." Rocks, too, can serve a similar nursing function. Lastly, it should be noted along these lines that tall cacti standing out in the open today may have long outlived the shrubs and trees that gave them their starts.

As the tiny globes of spines emerge, packrats and other animals feed on them, or ground squirrels and birds may dig them up while searching for other food. Heavy storms tear out their undeveloped root systems, or, contrariwise, a drought during the first year will shrivel them up. Unlike their parents, seedlings have little moisture storage capacity.

To risk speaking in anthropomorphic terms, seedling saguaros have a terribly risk-filled, life-and-death struggle. Few of them make it. During a hundred years of flowering and fruiting, a giant cactus spews out approximately forty million tiny seeds. If of these even two or three survive to

Avian dwellers of the saguaro forest. Top, left to right: white-winged dove eating ripe fruit; gilded flicker feeding its nesting young; black vulture. Bottom, left to right: elf owl resting in a palo verde tree; mockingbird perched on skeleton of cholla cactus; red-tailed hawk chick in its saguaro nest.

Animals of the saguaro forest. Top, left to right: kit fox approaching den of kangaroo rat; giant centipede; sidewinder rattlesnake. Bottom, left to right: kangaroo rat; javelina rubbing the scent gland of another; coatimundi, or chulo.

maturity, it would be a smashing success rate.

Yet in this respect, anthropomorphism is precisely the problem. Seeing the saguaro as "heroic," we lose sight of its context in the ebb and flow of desert life. Scientifically speaking, the saguaro is no more desirable or undesirable than any other plant or animal in the Sonoran Desert's ever-shifting dynamics. That it is visually arresting is important to the humans who see it as such; that it serves as one of the major linchpins holding the desert's ecological strands together is more subtly compelling.

For the saguaro is tenement as well as supermarket for many species. Favoring high points for lookouts, hawks, ospreys, and vultures cradle their roosts between the spiked Ys formed by the saguaro's arms. Carpenter birds, Gila woodpeckers, and flickers chip out holes in saguaro trunks. Burrowing deeply, they excavate nest cavities in the cactus pulp. Surrounded by bristling armament, these birds find refuge from egg-loving ringtail cats and Gila monsters, though coachwhip and gopher snakes slither up the green columns to eat the young. Still, such hideouts offer one of the best homes for birds. The living insulation keeps the temperature cooler than the outside air during the day and warmer through the night, and it retains humidity—prime considerations for rearing young in the blazing summer. One can hardly stroll in a cactus forest for more than a minute or two without spying the round entrances to the fortresses.

The arrangement is so popular that birds ill-equipped for excavation work compete for unused sites from past years. The list of these is long. Flycatchers, owls, finches, and sparrows all make use of abandoned cavities and rely on them for successful nesting.

Beyond food and shelter, in a land short on waterholes, saguaro cavities hold rain water blown in by violent thunderstorms, providing adventitious supplies for birds and for a mosquito that breeds solely in these uplifted, desert ponds.

Above the ground, insect larvae tunnel and feed inside the saguaro, while in their underground world cicadas cling to the roots. Pocket mice dig labyrinths among the root systems, and woodrats chew spiral galleries around the lower portion of the plants to obtain moisture. All this activity attracts the coyotes, foxes, and bobcats that live on the creatures that live from or in the saguaro.

Even the fallen giants continue to serve the community of flies, beetles, and lizards. These feast on or live in the bacteria-laden, rotten pulp of the plants. Finally come those excellent scavengers, the termites, who clean up death and prepare the way for new life.

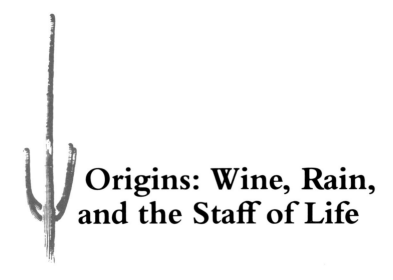

Origins: Wine, Rain, and the Staff of Life

Like most cacti, the saguaro has a leafless stem, is covered with spines, contains relatively little woody tissue, and carries its food-manufacturing chlorophyll in its skin. The result: the saguaro survives by storing water in its cells while the sun blazes on. Most remarkable for its size, the saguaro is the largest cactus within the United States, though the cardón of nearby Mexico grows taller and is more massive. All this not to demythologize the saguaro. Whatever science has learned, a good deal of telling mystery clings to it from the ages.

That a good portion of the mystery comes from confusion may seem strange for such an obvious and formerly useful plant. Yet often in human affairs, the most desired and familiar things cause the most furor precisely because they are at once multifaceted and integral to their surroundings. We easily pigeonhole less important matters.

For instance, the first documented encounter by an English speaker with the saguaro went like this:

A species of tree, which I had never seen before, here arrested my attention. It grows to the height of forty or fifty feet. The top is cone shaped, and almost without foliage. The bark resembles that of a prickly pear; and the body is covered with thorns. I have seen some three feet in diameter at the root, and throwing up twelve distinct shafts.

Snowfall on the saguaro forest. Such occasional cold spells limit the range of the giant cactus.

One would have difficulty, indeed, drawing a credible saguaro from the description.

Extraneous factors compound the confusion. In the 1820s, the above writer, young adventurer James Ohio Pattie, explored what is now Arizona when he journeyed west from New Mexico along the Gila River on a trapping expedition. For years scholars have debated the accuracy of his *Personal Narrative,* a sometimes bombastic account of his exploits. Not only does Pattie tend to get his dates wrong, sometimes by years, he obviously sprinkles the glitter of romance over his deeds as he recites a continuing drama of maidens, babies, and brave companions rescued from the maws of Fate. So when he mentions that the Southwest's javelina, a wild, piglike creature, has a navel on its back, readers might well cluck with further skepticism when they turn to his description of the strange cactus. Yet a closer, more patient, reading yields a different conclusion. As to the javelina, the mammal has an opening for a scent gland on the back, a feature understandably mistaken for a navel by a nonscientist.

Scientists today indeed consider the saguaro a tree, an arborescent cactus, to use the exact term. But the saguaro hardly seems cone shaped, and the tall cactus is not almost without foliage, it is bare of foliage. In the 1540s, Pedro de Castañeda, the chronicler of Francisco Vásquez de Coronado's expedition north from Mexico in search of gold and fame, called the saguaro "pitahaya." Along with other Spaniards, he was using the term learned from Indians elsewhere and applied to any giant cactus. Unfortunately, subsequent explorers continued to use pitahaya, much to the distress of modern botanists trying to pinpoint the saguaro's fluctuating range throughout history.

Yet naming is an unpredictable process. Somewhat curiously, English speakers eventually settled on "saguaro," itself a Spanish corruption of a local Indian word. But how to spell it? After reviewing the literature, one scholar listed, with chagrined humor, some of the variants: sahuaro, suwarrow, suwarro, suahara, zahuaro and zuwarro! Our only claim to victory along these lines is that, whatever the orthography, it is pronounced sa-wa′-ro—usually.

And, usually, where laymen founder, scientists rush in to put things in order. But even here a certain confusion reigns. In 1848 Dr. George Engelmann first described the Southwest's spiny giant and applied the scientific name *Cereus giganteus.* There matters rested for some sixty years. Then scientists at the New York Botanical Garden decided to change the name to *Carnegiea gigantea,* in honor of industrialist Andrew Carnegie, sponsor of botanical research in the desert. Cold science can breed its own versions of myth. Notes Frank S. Crosswhite:

There is a story, apocryphal indeed, but well known to many botanists, that the naming of the plant for Carnegie was a ploy to extract further funding from him for plant research. As the story is commonly told, Carnegie was asked to come to Arizona and look at the unusual plant which had been named for him. He obliged and was shown the great forests of Saguaro near Tucson which the Papago had depended on for hundreds of years. When told that these giants all now bore his name, he inquired innocently how such a characteristic and useful plant had evaded detection near a highly populated area of the state for so long. When the blushing botanists tried to explain how it had been scientifically necessary to change the name of the plant and that the cactus had been well known for many years, Carnegie reportedly became disgusted with a kind of science that could change well-established names of plants to cater to a wealthy person.

Whatever the motives of the blushing two, in all likelihood quite honorable, they further propagated the perplexity. The nation's largest and most famous cactus, praised in song and prominent in western cartoons, remains in nomenclatural limbo. Wherever people who care about such things gather, they remain at loggerheads over the propriety of *Cereus giganteus* and *Carnegiea gigantea.* Ironically, unlike Pedro de Castañeda, we now know what we are talking about but remain divided as to what to call it.

Admittedly, all this is a fairly minor matter, a mere footnote, though a telling one, in the annals of science. Today, botanists concentrate their energies on more important aspects of the giant cactus: its shifting range, its role in the ecological web, its chances for survival in the face of the overwhelming postindustrial invasion. To the Indians of former times, however, the saguaro came close to meaning life itself. For the saguaro and the Papago Indians inhabit a harsh land. Marveling at the Indians' ability to live under the conditions, Frank Russell, an early anthropologist, called it an "all but hopeless environment," a "vast territory of cactus-covered plains, here and there interrupted by up-thrust barren peaks that, with striking outlines, form good landmarks and yet offer little to those that hunger and are athirst."

It is not an exaggeration for dramatic effect to apply the words "rugged" and "harsh" to this region of extremes. In this corrugated basin-and-range area, peaks studded with cold-weather species of pine and fir alternate with broad, desert valleys miles across. Here, where the daytime temperatures in summer regularly soar over 100° F, rainfall is sparse, sometimes no more than a few inches each year. In consequence, where most needed, the perennial streams are few, and the numerous arroyos, or dry washes, run only during the rainy seasons of winter and summer. Yet what surprises the visitor expecting camels plodding dutifully over sand dunes is the apparent lushness of vegetation and the wealth of animal life. The explanation is simply stated though complex in its details: life thrives in the "hostile" desert because it has adapted over the eons to the conditions by husbanding water, conserving energy, waiting patiently for the next rain.

In such a landscape, the Indians lived, sometimes comfortably, at others just barely, because like all the other living things around them, they used well what little nature gave them. Nature itself, in fact, could be an abundant storehouse. The high elevations offered the skins and meat of white-tailed deer and wild sheep, while out on the flats the cautious hunter found mule deer, javelina, and a variety of smaller game. As to vegetation, ingeniously the Indians turned to their own ends what may strike us as far too spiky and desiccated to be of much good.

The night-blooming cereus seems a weird, snakelike plant, its single virtue a brief blooming of large, white, and pungent flowers on hot desert nights. Indians followed the sweet smell to the plant, there to dig beneath the spiny affair for its huge tuber, weighing up to fifty pounds and a delicacy when roasted. Various cacti provided not only food and moisture but fishhooks and, in modern times, even phonograph needles! Indians turned the roots of lotebush into shampoo and the seeds of wild mustard into a medicine for sore eyes. The mesquite tree bears beans for food, and its hardy wood serves splendidly for cooking-fires, tools, and construction materials. Nature not only was pharmacopoeia, clothing store, and supermarket, but a sewing basket, offering thorns for needles and fiber for thread. The list could go on almost indefinitely. As wondrous as all this may appear to those of us dependent on manufactured goods trucked in from afar, in context there is nothing exceptional about such desert lore. Tribes that did not learn to use while conserving their resources did not survive; those that did lived on.

In this sense, there is little particularly romantic about Indians' "oneness with nature." Native Americans attuned themselves to the environment as a method of survival, and how they lived had everything to do with where they lived. In areas around modern-day Phoenix, Arizona, for example, the Indians lived for hundreds of years primarily as farmers. Turning to good advantage the rivers carrying snowmelt across the desert from mountains miles away, they dug irrigation ditches out to their fields of corn, beans, squash, and cotton. Spanish missionaries later introduced European crops, adding a welcome variety—wheat, oats, peaches, carrots, onions, barley, and peas. For the most part, theirs was a sedentary life, wedded to the watercourses that gave them sustenance. Saguaros grew around them, but the cacti were not essential to their livelihood.

Speeding along a highway to the south, the traveler now notices little difference in the homeland of their cousins, the Papagos.* Here, apparently, are the same mountains, vitreous in the heat of noon, and the same desert vegetation in the broad valleys between them. But the Papago land has few year-round streams. Planted in narrow flood plains, their irrigated crops were tentative at best as survival insurance against hunger. Perhaps they flourished one year but did poorly the next. The circumstances forced Papagos into a hunter-gatherer pattern, an existence gleaned largely from wild products. In this scheme, the plentiful saguaro played an essential role, providing the Papagos not only with shelter and tools but offering food just at the critical time, early summer when other plants had not yet ripened. Lastly, the giant cacti gave them the ability, so it was belived, to control the very rains for their domesticated crops that could mean feast or famine for the rest of the year.

In their houses, some Papagos still use the woody ribs of saguaros in doors, walls, ceilings, shelves, and furniture. They further use, or once did, the long sticks in building cradles, carrying baskets, small animal traps, and bird cages. Saguaro products served in the tanning and weaving processes and as

*The Papago people recently changed their name to Tohono O'Odham. The former name means "Bean Eaters," the latter, "The People."

tools, especially in making bow strings and fire. In a society of scarcity, even the saguaro spines were not totally discarded but sometimes found employ as needles in tattooing. Papago children played with saguaro doll houses and saguaro noisemakers, while adults also depended on the cactus for items in games and for wood to carve religious fetishes. Medicinally, pieces of the cactus served as pain killers. A saguaro gruel supposedly helped mother's milk flow, while the ever-handy ribs made convenient splints. One hardly could turn around in a Papago community without seeing some evidence of the great cactus woven, spliced, nailed, or tied into Papago life.

To a far lesser degree this is true today, for manufactured goods have replaced most handmade items. Although, like Anglos, the Papagos now shop in supermarkets for most of their food, the saguaro was central through the centuries to their diet, and the Indians still feel the tug. How else could it be when their very calendar begins with *Hahshani Mashad,* the month of June, and the time of the saguaro fruit harvest?

Back in the days before the Anglos' influx disrupted ancient ways, the Papagos moved around a good deal in search of food. They spent summers out on the hot lowlands, tending their fields while anxiously eyeing the southeast skies for signs of rain. Winter took them up into the mountains, where canyon springs held the all-important water. Late May or June found them moving down into cactus camps, hereditary territories in the foothills among thick saguaro stands. This marked a happy time, for the prospect of harvesting the tasty fruit tinged the work with aspects of a vacation. It meant the first fresh, juicy food after the long winter fast, combined with a gathering of dispersed families. All this occasioned a feast accompanied by songs, good talk, and thanksgiving, sometimes by a baptism with cactus juice.

Today, as did their grandparents, the Indians fashion a long pole of saguaro ribs to get at the plum-sized, dark red fruit growing high on the tips of the cactus. They finish the simple tool with two crosspieces, one near the end and another in the middle for the lower fruit. Striding out early in the morning before the heat of day, the Indians work their way back to camp, some of them hooking down the fruit with their poles, while others collect it in baskets. A flick of a knife or even a fingernail specially groomed long for the purpose separates the husk from the valuable pulp. Even the discarded part, however, may serve a purpose. Some Papagos place it on the ground fresh side up as a prayer for good summer rains.

Meanwhile, people back at the saguaro camp have been busy gathering wood, for although Papagos relish fresh fruit, they process most of it into a variety of syrups, jams, flour, gruels, candy, and a

The saguaro fruit harvest. Left: hooking the saguaro fruit. Above, left to right: the succulent saguaro fruit pod; at camp, the fruit is soaked in a pail before being cooked in a large pot; a typical Papago saguaro harvest camp. Bottom, left to right: the cooked juice is poured through a piece of cloth to filter out the seeds; the filtered juice is cooked over a hot fire for several hours until thickened into a syrup; glass jars are filled with the boiled-down syrup.

dried pulp eaten either raw or mixed with other foods. In the past, before the days of refrigeration and grocery stores, they set aside a large portion to help tide them over through the lean winter months. Usually, whatever the end purpose today, preparing the fruit entails cleaning, soaking, and boiling. Though the Indians enjoy the tiny seeds when made into cakes, they also feed them to their chickens.

What was corn, wheat, flour, sugar, and honey to most of our ancestors, the saguaro was to the Papagos. Beyond domestic use, saguaro products once served an economic function. Lacking most of the desired trade items, the desert tribe carried balls of saguaro pulp and clay jars of syrup on their backs to exchange for grain, pottery, and abalone shell with neighboring tribes. But most of all, the saguaro was a nutritious and flavorful staple—tasting something like raspberries—rich in oils, a source of B vitamins, and a reward for a sweet tooth. In days past, the Indians so enjoyed themselves during the several weeks of gathering time, gorging themselves on the fruit, that an early traveler reported with some amazement:

> They can eat as much of it as they please, and with some this food agrees so well that they become corpulent during that period, and for this reason I was sometimes unable to recognize at first sight individuals otherwise perfectly familiar to me, who visited me after having fed three or four weeks on these pitahayas.

After harvest, the Papagos faced many lean weeks before other desert plants bore fruit and even more before their small patches of corn and beans matured—if there was rain enough to mature the crops. Further, there is some evidence that Papago metabolism adjusted over the years to just such cycles of feast and famine.

As to feasts, even today one might chance upon an apparently bizarre scene on the desert, though the Papagos do not advertise it to the general public. Among a large gathering of Indians in a cleared, isolated place, people stumble around as if drunk. Now and then, some of them leave the crowd to vomit. They are indeed drunk—or at least quite tipsy—but what might strike the outsider armed only with preconceptions as weird, if not outrageous, behavior, actually is an ancient and complex religious event. In former times, it had everything to do with the Papagos' survival. Not strangely, the ubiquitous, useful saguaro stands at its very center.

The women have not turned all the cactus fruit into jams and syrup and cakes. They set some of the juice aside in large ollas, or pots, and place them in a house built for a special purpose. Outside it, the people dance, helping the liquid along as it ferments into wine. They are here, however, not only to have a once-a-year bash with an alcoholic beverage, and not simply to gather a normally scattered people for talk and good times before they disperse in family groups to spend the summer working in their fields.

This ceremony, as with most native American ceremonies across the continent, serves multiple ends. The major purpose is practical: to saturate their bodies with wine, as rain saturates the skies and earth. In this magical way they encourage the precious summer rains that will make their fields thrive.

Because the event has multiple purposes, so the accompanying emotions are equally varied. In days past, the shaman, or head religious man, of each village went through exacting rites in overseeing the fermentation. He cleaned the wine jars by blowing smoke into them. He gathered contributions of juice from the households and sent men out to bring back water from special springs to use in the mixing. The whole process confirmed the seriousness of the occasion and the delicate position of men in their relationships with nature. Dancers chanted sacred songs to the cardinal directions; runners sped off to deliver elaborate and carefully memorized speeches of invitation to neighboring villages.

Once the wine was ready and served up, the people listened to more tales of the ancient ways, all confirming the role of the Papagos in the pattern of the cosmos. Elders reminded them they were not to overindulge; they were to avoid quarrels that would mix liquor with blood. In days gone, the feast produced its share of rampant revelry, sexual license, and, sometimes, violence. Strutters painted the soles of their feet red so that the striking color would be their final encore after toppling over. In this atmosphere, women and children from modest families fled to the housetops for security from the dizzy commotion in the village below.

In the past, too, the period of celebration brought outside danger. Apaches came down out of the mountains to lurk around the festivities, waiting to attack their enemies if they let their guard down too far. Also of old, the ceremony performed across the farflung Papago land had local variations. Today, these are not as elaborate as before. Still, in a far more significant way than most Anglos celebrate the New Year, Papagos look forward to the turns of the seasons ahead. They may drink their sacred liquor from some old coffee cup or tin can lying about, and to an Anglo, the surroundings appear decidedly uncathedral-like; but they are intensely engaged because each person feels directly involved in an activity having everything to do with the group's welfare. Enjoying themselves, constantly socializing as the ritual unfolds, they grow increasingly exhilarated, not only from the wine, but from knowing that their joy helps along the coming of rain. Anthropologist Ruth Underhill notes of the drinkers staggering off to vomit, an act that would cause much clucking in most white churches:

This is recognized as a ceremonial feature, and people say with pleasure, pointing out a man so affected: "Look, he is throwing up the clouds." The regular procedure during the twenty-four hours of feast is to drink, vomit, sleep, and drink again, until the result is a thorough purging.

And then, the fun over, confident of the efficacy of their pleasures, the Papagos settle down to the weeks of fieldwork ahead.

This part of the Southwest receives almost all of its summer rainfall from moisture blowing up out of the Gulf of Mexico hundreds of miles away to the southeast. Eyes turn in that direction as the season begins by stages. Almost every day the bruise of clouds grows larger and purpler—only to dissipate in the dry desert air. But the buildup continues, and one July day the landscape darkens as if night were coming on. The saguaros stand out in eerie half-light. The farmer leans on his hoe, watching the drama of a massive storm rolling over the mountains toward his fields, watches the first heavy drops splatter in the dust around his feet.

In contrast to the Papagos' relationship to earth and rain, the larger technological society crowding around and sometimes crowding upon the Indian homelands and traditions has little physical or spiritual connection to the saguaro and, hence, feels no such tenderness toward it. Builders in the Southwest move in their earth-moving machines, and in a blink shave off complex communities of desert plants as a prelude to expanding suburbia. On weekends, Anglos prowl the desert in chrome-plated, four-wheel-drive contraptions, wantonly blazing away at the humanlike targets of the tall cacti. The bullet wounds cause black scars readily seen on saguaros in southern Arizona wherever highways or dirt tracks make them accessible to the public.

At least once, patient nature in return dealt out poetic justice, as reported in the Tucson *Star,* February 6, 1982:

Saguaro's last act avenges its death

PHOENIX (AP)—A 27-year-old Phoenix man was killed when a saguaro cactus he shot fell on him, authorities said. Maricopa County sheriff's deputies said David M. Grundman fired a shotgun at least two times at a 27-foot cactus. The shots caused a 23-foot section of the cactus to fall and crush Grundman, deputies said. Deputies said Grundman already had felled one saguaro. Destruction of cacti is a misdemeanor under Arizona law. Grundman and a friend, Joseph Suchochi, were in a desert area north of here when the incident occurred Thursday afternoon, deputies said.

Yet Anglos have not been entirely insensitive to the giant cactus. Urged by citizens concerned about the natural heritage passed on to their children, the

Road building is a traumatic disruption to the saguaro and its habitat.

government has set aside large reserves, such as Saguaro National Monument near Tucson. These preserve the habitat not only for the tall cactus but for all the native plants and animals, while offering hundreds of square miles of wild peaks and rugged canyons in which hikers can lose themselves to their hearts' content.

On a lighter note concerning appreciation, it is almost impossible not to make a connection between the single-trunked, many-armed cactus and the human outline. One nineteenth-century German traveler through the saguaro forest enthused that, "Some look like petrified giants stretching out their arms in speechless pain, and others stand like lonely sentinels keeping their dreary watch on the edge of precipices." That sentinel image stuck, becoming a cliché dear to southwestern versifiers. It is reported, however, that one recent Girl Scout leader embellished the analogy to livelier effect. Backpacking with her troop in the saguaro forest, she began making up idle tales for her young charges as they sat around their campfire one night. Around them, the flames swayed over the giants. If the girls happened to wake up in the night, she mentioned, they might listen carefully. Sometimes, when the moon is just right, she whispered, these monsters grow weary of holding out their arms and take a bit of exercise, one deliberate step at a time. On these occasions, one can hear the dull thunder of their footsteps echoing over the lava hills.

In the morning wide-eyed girls danced around her, insisting that they not only heard but saw the saguaros about the camp taking their surreptitious constitutionals.

"Looking for My Mother"

"Long before the coming of Coronado, Papago Indians had regard for the saguaro. It was their food and drink, their shelter and religion; no other single factor was so important to the tribe's welfare. While four centuries have brought many changes to the Indian mode of life, the cactus still ranks high in their esteem."* This respect engenders a human identification impossible for most Anglos. In his recent book, *The Desert Smells Like Rain,* Gary Nabhan catches a vignette during the saguaro harvest season:

I was startled when I came upon some Papago who clearly take heed to the "human"

vulnerability of saguaros. I heard a young city boy ask an elderly Papago woman if, lacking a harvesting pole, one could ever collect fruit off the tall cacti by throwing rocks at the tops to knock the fruit down.

"NO!" Marquita replied with a strain of horror in her voice. "The saguaros—they are Indians too. You don't EVER throw ANYTHING at them. If you hit them in the head with rocks you could kill them. You don't ever stick anything sharp into their skin either, or they will just dry up and die. You don't do anything to hurt them. They are Indians."

So deep does the feeling go that Papago road builders have been known to swerve a new project around one of their fruit-bearers.

---

*Nell Murbarger, student of the Papagos and the role of the saguaro in contributing food, drink, and shelter.

Evening in the saguaro forest

Reverence for the aspects of the natural world on which their lives depend typifies traditional peoples wherever they live. It goes far beyond showing itself in the reprimands of a Marquita to a brash youngster. Tribal myths tell individuals who they are and how to act. Myths express the ways the members of a group see themselves as they try to make sense of a complex and often hostile world. Not all of this is dour, however. Quite typical of native American lore, the land of tales is a colorful and magical place, where animals speak and things change shape, where unexpected events have unexpected outcomes. Often it is a world of humor. Today the mockingbird babbles in the treetops because he once got drunk. He has been talking away ever since.

Quite typically also, then, the saguaro plays a large role in traditional stories the Papago tell about their desert and how the things in it and their way of life came to be. Much of this explains how the Indians received the life-giving gifts of nature, primary among them the saguaro. For instance, this dreamlike tale told by Susie Ignacio Enos, describes a girl who is transformed into a saguaro and helped by a bird:

●

Once there was a little Aw'awtam baby who lived with her mother and father, and her name was Sugu-ik Oof. She was called this by her mother because she was always happy.

Now it happened that when she was about a year old, her father, who was very fond of her, died. Before he closed his eyes in sweet sleep, he said to his woman, "Bring Sugu-ik Oof to me." When the mother had brought the child to him he said, "The Wise Man of the Aw'awtam has appeared to me as a young man, but here I lie and am about to leave you, so it is that since Sugu-ik Oof is a girl she will be different from woman. She will be great as her father should have been. She will live forever to the end of times. She will be known by races of people from far and near. She will be queen of the Taw haw naw Juwut (desert lands). Generations of Aw'awtam will be saved from starvation because of her and her family."

When the young father had spoken he closed his eyes in peaceful sleep.

Years came and years went and Sugu-ik Oof grew in beauty and in size. When she was ten years old she could not stand the loneliness of her life any longer, for her mother had to go into the village nearby to get food for them both, every day. The little girl was left all alone day after day.

So one day Sugu-ik Oof said, "I will go after my mother for I am very lonely at home by myself."

The little girl set out on her journey, although she knew not where her mother could be. Off in the distance were the Gihau mountains.

"My mother pointed out the Gihau mountains to me so I shall go toward them; maybe that is where she is."

Sugu-ik Oof walked and walked for a long time before she saw any kind of life. In the distance a coyote approached her.

"My friend, Coyote, will you tell me where I can find my mother?"

The Coyote, with sly eyes, said: "Yes, Sugu-ik Oof, I will tell you where your mother is if you will give me one of those gourds you are carrying."

The little girl handed one of the gourds to the Coyote very anxiously.

"Over the Gihau mountains you will find a village and there you will also find your mother," the Coyote said and went on his way.

Sugu-ik Oof walked on and pretty soon she met a rabbit.

"Be very kind to me and tell me where I can find my mother," the little girl again asked of the animal.

So the rabbit looked at the gourds very fondly and said, "If you will give me one of those gourds you are carrying, I will tell you where your mother is."

"My mother gave these to me and I like them very much, but I will give you one gladly." So she did.

"Beyond the Gihau mountains you will find your mother in the village," said the little rabbit and ran off on his way.

Again Sugu-ik Oof started walking. By now she was very tired and hungry. The mountains to which she was walking seemed so near and yet when she walked toward them they seemed to get farther away.

When the sun was turning toward the west, she met a little bird who spoke very kindly to her.

"Sugu-ik Oof, why are you this far away from your home? It is getting late and shadowy. Your mother will be unhappy if she doesn't find you at home."

"Little Gray Bird, I am looking for my mother. Can you tell me where she is? I will give you one of these gourds if you tell me," the little girl replied.

"I shall be very happy to tell you. I will even show you the way to the village, but I cannot take you clear to the village for there the little Aw'awtam boys shoot arrows at me and throw rocks," the little gray bird told Sugu-ik Oof.

"I am very happy because you have been very kind to me. Some day you will be rewarded for what you will do for me this day. Shall we go?"

The two traveled on and on, over rocks and down slopes they went; the little bird ever flying by Sugu-ik Oof showing her the way. When the shadows of evening began to fall they had climbed over the mountain.

The little gray bird said, "Now, I will leave you but from here you can see the village where your mother is." And so the bird left Sugu-ik Oof.

Sugu-ik Oof walked down to the village where she saw a group of children playing.

She went up to them and asked, "Can you tell me where I can find my mother?"

The children didn't seem to hear her for they did not answer her. She waited for their answer and once more asked, but they still would not answer her.

Pretty soon Sugu-ik Oof began to chant a song. The children still did not pay any attention to her. She chanted on and on. When finally the children looked at her, they saw that she had sunk halfway into the ground. At this the children became frightened. They all started screaming for help. Some of them rushed to the place where they knew her mother worked and they quickly called her. In the meantime one of the little boys pulled her by her hair thinking he could save her, but it was no use, for he only pulled the hair off from the top of her head. They could not help her for she had sunk deep into the ground and was no more. Her mother came running but she was too late. No amount of tears could bring the little girl to the surface of the earth.

Sugu-ik Oof's mother used to visit the place where her daughter had sunk and she would place food and water there because she believed that her daughter's spirit still lingered there.

A whole year went by, and it was the same season the next year. When Sugu-ik Oof's mother went to place the food and water at the spot, she noticed a queer plant had come up in the very place where her daughter had gone. So the woman began to water the plant and to take care of it, until it grew into a tall and stately plant. No one had ever seen any plant like it. People came from far and near to see this peculiar plant that had grown where Sugu-ik Oof had sunk.

"What use can a plant like that be to us? It has too many stickers and we cannot eat it," many of them were saying.

Years went by and Sugu-ik Oof's mother had grown old and feeble but she was faithful to her plant and daughter for she still took care of it. One day after the coldest time of the year had just passed, when it had grown so high that it soared higher than the highest plant in the desert, the stately plant began to show signs of budding. The whole village was excited over this but Sugu-ik Oof's mother was even more excited.

The buds grew and grew and when the desert flowers just began to peep out of the ground, the queer plant bloomed forth into a beautiful white flower. After the flower had gone, a fruit formed and this grew from a green to a red fruit.

One day when the flowers had gone to seed and the desert began to swell with the summer heat, the fruit of the stately plant burst, showing forth a scarlet red. When it had fallen to the ground Sugu-ik Oof's mother ate some of it after she had seen the birds eat it when it was still on the plant. She tasted that the fruit was delicious and so she gave some to the other people and they too liked it.

Every year after that the *hash'an,* as the people began to call it, bore the same flower and fruit. The birds liked it so well that as soon as the fruit began

to show signs of ripening they would gather on the plant and eat of it. The Aw'awtams liked it too and so were angry at the birds who could reach it more easily, so the children began to shoot the birds with their bows and arrows and they threw rocks at the birds to chase them away.

One day at the time of the ripening of the fruit the children ran out to get at the fruit before the birds, but they found that the *hash'an* had disappeared. With great excitement they ran and told the other people, who ran out to see if what the children were telling them was true. They were sad for their delicious fruit had disappeared. They became alarmed, for they believed it to be the work of evil spirits. Thereupon a great council was called at which were present the wisest men of the villages far and near. Smart medicine men were there too to give their aid.

After many nights of council they decided that everybody that was able, man, beast and fowl, should go out and hunt for the plant.

The hunt started. The birds hunted in the air as far and as high as they could go; the animals that live mostly under the ground dug holes—deeper than they had ever dug before to hunt there; the men and animals alike hunted all over on the surface of the earth. The search lasted for days and days until one day they began to come back one or more at a time and began to unfold strange adventures concerning their search. They had seen many strange and beau-

Spring is the season for blooming. Here in the foothills of southern Arizona, palo verdes (yellow blossoms) and saguaros (white blossoms) put forth flowers at the same time.

tiful things but they had not seen their plant.

Finally all but one returned. The one who had not returned was a small gray bird who had only one eye. The Aw'awtam children had hit him in the eye when he was trying to eat some of the *hash'an* fruit.

Those who were gathered there began to make fun of the little bird. "Oh, he has only one eye and he probably got lost somewhere. Anyway he couldn't have found the plant with his one eye." On they talked in this way about the poor little gray bird.

Several days went by. One day the keen-eyed eagle said that he saw a tiny speck in the sky in the very far distance. Maybe it could be their friend the little gray bird. And so it was. It wasn't long when the little bird came all tired out and hungry for he came so fast to tell the news. He told that he had found the *hash'an* high on the west side of a high mountain.

A group of people got together and made their way into the mountains to see for themselves.

When they reached the place, sure enough the *hash'an* was there where the little gray bird directed them. The *hash'an* was asked why it had tried to disappear and this is what it said:

"Years and years ago when I was a little girl in the form of an Aw'awtam, I was hunting for my mother and a tiny gray bird like the one who found me was kind to me and willingly helped me to the village where my mother was. It was then that I promised that one day the little bird would be rewarded for his

kindness. Now I am a *hash'an,* I can help the birds. They will be the first to eat of my fruit when it is ripe. When I stood near the village the Aw'awtam children used to shoot and throw rocks at the birds when they tried to eat of my fruit, but I didn't like that. If the birds couldn't eat of my fruit then no one else could."

So it was decided there between men, beast, and fowl that all should eat of the fruit of the *hash'an,* the tallest and most stately plant on the desert of this country.

We should remember that Papagos do not spin traditional tales as we might, making up parts of a story or adding color as we go along. The stories are too sacred, too integral to a culture delicately balanced in the universe, to be so treated. Nevertheless, Papagos have been telling stories for hundreds of years in dozens of isolated communities. Though the basic ideas behind them remain quite constant, inevitably, over time and space, changes slip in and details evolve into charming variations. In this tale collected by Harold Bell Wright, it is interesting to note not only the continuing help of animals—and sometimes their obstructive tricks—but the warning against injuring saguaros. As with so many blessings in human affairs, the saguaro wine becomes too much of a good thing, so it is taken away. This is another caution. The lost gift, once returned,

becomes ever more precious, ever more respected.

•

In a certain village there lived an Indian girl who did not wish to marry.

This girl would never listen to the women when they tried to teach her things. All she wanted was to play taw-kah all day long. Taw-kah is a game very much like shinny or hockey. It is played by the Indian women. And this girl was a very swift runner and such a good player that the people who bet on her would always win.

Finally the girl's father and mother grew tired of her playing taw-kah all the time. So they married their daughter to a good man. And they made her a fine new house where she and her husband went to live.

But very soon the new house was dirty. There were never any beans ready for the husband when he came from the fields. Again the young woman was spending all her time playing taw-kah.

At last the young woman's husband grew tired of the dirty house and of having to work all day and then prepare his own supper. So he went away and left her.

After a while the woman had a son.

The older women of the village thought that surely now this young mother would stop running about playing her favorite game. But the young

Sunset over the saguaro forest

mother filled a big gourd with milk and put it beside her baby. Then she fixed her hair with a bright feather and put beads around her neck and went out to play taw-kah just as she had always done.

But the women of the village did not like to play taw-kah with any one who always won. And they did not like to think of this young women's baby alone, crying. So the women of the kee-him would not play with this young mother. And she went over the mountain to another village where the people did not know her. She played taw-kah with the women of this kee-him and always won. And after a time she left this kee-him and went to a more distant village where she continued to win.

Now when this woman's child was left alone he began to grow. He looked around the room and saw all the beads and feathers and baskets and ollas which his mother had won playing taw-kah. He saw the gourd full of milk. He drank the milk and grew larger and larger.

Then he arose and put a feather in his hair and took all the bright things which his mother had won and went outside the house.

When the women of the kee-him saw the boy they said: "Here is the boy without a mother."

But the boy said: "No, I have a mother but she is playing taw-kah somewhere. Can you tell me where?"

The women could only tell the boy that his mother had gone away over the mountains.

So the boy started after his mother. He carried all

the things which his mother had won from the other women. And as he traveled he gave these things to the people who gave him food. But the way was very long.

After a time the boy came to some cultivated fields which belonged to a man who had a great deal of wheat and was busy clearing out the weeds. And the boy asked his question again: Where could he find his mother?

And the man answered that he knew the boy's mother well. He said she was always playing taw-kah and always winning, that she always wore a bright feather in her hair, and that she was in a kee-him on the other side of a black mountain.

So the boy started up the black mountain.

The trail was long. Sometimes Hoo-e-chut—a lizard—would stop long enough to laugh at him. Sometimes Oo-oo-fick—the birds—would fly very near and tell him not to be troubled.

Finally the boy reached the top of the black mountain and saw the village which lay below. There were some women playing taw-kah. And as the boy watched he saw a young woman who always won. And in this one's hair there was a bright feather.

The boy went on down the mountain to the village. But he did not go into the kee-him. He stopped in an arroyo near where there were children playing. And the boy asked one of the children to go to the woman who was playing taw-kah—the woman who had a bright feather in her hair and who

always won—and tell her that her son had come and wanted to see his mother.

The child carried the boy's message.

But the woman answered: "I will come as soon as I win this game."

The boy waited some time. Then he sent another child with a message to his mother.

But the boy's mother had not yet won that game of taw-kah so she sent the same message back that she would surely come when she had won.

Again the boy waited. And the boy was hungry. So he sent the third child to his mother begging her to please come soon because he wanted to see her.

But this game of taw-kah was very long and the boy's mother had to run very swiftly. So she answered the messenger, "Yes, yes, tell him I will come when I win this game."

When the third messenger returned the boy was angry. And he asked the children to help him find the hole of Hee-ah-e—a tarantula.

When they had found a tarantula's hole the boy asked the children to help him sing.

So ah-ah-lee—the children—formed a ring around the boy and began to dance and sing. And as the boy and the children danced and sang the boy sank into the tarantula's hole. With the first song he sank as far as his knees. He asked the children to sing louder and to dance harder. And as they circled around him singing and dancing he kept on sinking into the ground.

When only the boy's shoulders were above ground, one of the children ran to his mother, who was still playing taw-kah, and told her she must come quickly because the strange boy was almost buried in the tarantula's hole.

The mother dropped her taw-kah stick and ran as fast as she could. But the sun was in the mother's eyes and she could not see to go very swiftly. And when she reached the arroyo there was nothing to be seen but a bright feather sticking out of a tarantula hole. And the sand was closing around the feather.

Then the woman began to cry.

And Pahn — the coyote — who was passing, came to see what all the noise was about.

The mother told Pahn that her son had just been buried in the tarantula's hole. And she asked Coyote to help her dig her son out of the ground.

Coyote told the woman he thought he could get her son out. And he began to dig. And Pahn found that the boy was not very deep in the ground.

Now, Coyote was hungry with all this work, and he didn't see why he should take this boy to a mother who had never done anything for her son. So he ate the boy. When the bones were well cleaned Coyote took them out of the hole and gave them to the woman with the bright feather. And Pahn said to the woman: "Some one must have eaten your son. This is all I could find."

The woman with the bright feather looked at the bones. But the bones of her son were not very bright and so she had no use for them. She told Pahn to bury them again, which Coyote did.

Four days later something green came out of the ground on the spot where the boy's bones were buried. In four more days this green thing was a baby sahuaro — ah-lee-choom hah-shahn. And this was the first sahuaro, or giant cactus, in all the world.

This giant cactus was a very strange thing. It was just a tall, thick, soft, green thing growing out of the ground.

All the Indians and all the animals came to look at it. Ah-ah-lee — the children — played around it and stuck sticks into it. This hurt Hah-shahn and he put out long sharp needles for protection so the children could not touch him. Then ah-ah-lee took their bows and arrows and shot at Hah-shahn. This made Giant Cactus very angry. He sank into the ground and went away where no one could find him and he could live in peace.

After Giant Cactus disappeared the people were sorry and began looking for him. They hunted over all the mountains near the village. They asked all the animals and birds to help them.

After a very long time Ha-vahn-e — the crow — wandered over Kee-ho Toahk, which means Burden Basket Mountain. And Crow told the people that he had seen a very large cactus where there was nothing but rocks and where no animals nor Indians ever hunted.

Kooh — the chief — called a council of all the

animals and the people. And Kooh told the people to prepare four large round baskets. Then Chief gave Crow orders to fly back to the giant cactus. And Kooh told Ha-vahn-e what he should do when he arrived.

When Crow reached the giant cactus he found the top of Hah-shahn covered with fruit. The fruit was red and large and full of juice and sugar. Crow gathered the fruit as Chief had told him to do, and flew slowly back to the village.

The people were waiting.

And Crow put hah-shahn pah-hee-tahch—the cactus fruit—into ollas, which are large jars, and which were filled with water. Chief placed the ollas on the fire and from sunrise to sunset the fruit was kept boiling.

For four days this syrup—see-toe-ly—was cooked. Then Kooh told all the people to prepare for a special feast—nah-vite ee—which is wine feast, or wine drinking. They were to have a wine which they had never had before.

Oo-oo-fick—the birds—were the quickest to get ready for the feast. They came dressed in red and black and yellow. Some of the smaller ones were all in blue. Then Kaw-koy'—the rattlesnake—came crawling in. And Kaw-koy' was all painted in very brilliant colors.

The birds did not like it because Rattlesnake was

An unusual, twisting formation for a saguaro

painted so bright. They gossiped and scolded and were jealous. Rattlesnake heard Oo-oo-fick talking and rolled himself in some ashes. And that is why, even in these days, you will find the skin of Kaw-koy' marked with gray. The gray markings are where the ashes were caked onto his new paint.

Choo-ah-tuck—the Gila monster—gathered bright pebbles and made himself a very beautiful coat. And Gila monster's beautiful coat was very hard. You can see it today because he is still wearing it.

So all the people—Indians and animals and birds—gathered around and drank nah-vite—the wine. And nah-vite was very strong. It made some sing. Others it put to sleep. Others were sick.

Choo-hook Neu-putt—the nighthawk—who was dressed up in gray and yellow, did not wish to spoil his breast feathers so he brought a stick of cane to drink through.

All the Indian girls thought this very wonderful of Choo-hook Neu-putt and he received a great deal of attention. And Saw-aw—the grasshopper—who had borrowed Tawk-e-toot—the spider's web—and with it had made himself beautiful new wings, was very jealous to see the attention given Nighthawk. Grasshopper felt he must do something to make the people notice him. So he pulled off one of his hind legs and stuck it on his head.

When Nighthawk saw Grasshopper with his new headdress he laughed and laughed and laughed until he could not stop laughing. He laughed so hard he split his mouth. And it is that way even to this very day. That is why the nighthawk never flies in the daytime. His mouth is so big and white and ugly that he has to fly at night so people will not see him. And that is why he darts past you so quickly in the evening.

Often in these days, too, you will see Saw-aw jumping around without a leg.

After a while, as they kept on drinking nah-vite, the birds all began to fight. They pulled each other's feathers. And some had bloody heads—just as you see them today.

When Kooh saw the fighting and the bloody feathers, he decided that there should be no more wine feasts or wine drinking like that. So when the wine was gone, Kooh very carefully gathered all the seeds of the giant cactus fruit and he called a messenger to take the seeds away off toward the rising sun.

The people watched Chief's messenger take the seeds away off into a strange country and they did not like it. So they held a council. And Coyote said he would go after Chief's messenger.

Coyote traveled very fast. He circled around the one who was carrying the seed and came back so that when he met the messenger it appeared that he was coming from the opposite direction. Pahn greeted Messenger and asked what it was that Messenger carried in his hands.

The one who had the sahuaro seed answered: "It is something Kooh wants me to carry away off."

Coyote said: "Let me see."

Messenger said: "No, that is impossible."

But Coyote begged: "Just one little look."

At last, after much coaxing, Pahn persuaded Messenger to open one finger of the hand which held the seed.

Then Coyote complained that he could not see enough and begged Messenger to open one more finger.

And so, little by little, Messenger's hand was opened.

Then, suddenly, Coyote struck Messenger's hand and the seeds of Hah-shahn—the giant cactus—flew into the air.

Huh-wuh-le—the wind—was coming from the north. Huh-wuh-le caught up the seeds and carried them high over the mountaintops and scattered them all over the south side of the mountains.

And this is why the sahuaro, or giant cactus, still grows in the Land of the Desert People. This is why Hah-shahn is always on the southern slopes of the mountains.

Ever since that time, once each year, the Indians have held nah-vite ee—the feast of the cactus wine.

The Future:
The Great
Saguaro Debate

In 1969 writer Joseph Wood Krutch talked to a newspaper reporter about the saguaro. A few years earlier, the former Columbia University literature professor and drama critic for *The Nation* had retired early and moved to the saguaro forest. He explained the abrupt change: "I came for three reasons: to get away from New York and the crowds, to get air I could breathe and for the natural beauty of the desert and its wildlife." Krutch did not rest at the oars. Instead, his newfound ease and peace of mind inspired him in a different direction. More and more fascinated by the delicate interplay of flora and fauna surrounding his desert home, he turned out a book, sometimes two, each year, celebrating the natural complexities of the arid landscape. Soon the

former scholar and critic had won a second fame as the grand old man of desert literature.

But now something was going wrong with the desert, at least with the saguaros standing at the center of its ecological web. The saguaros were dying.

At this time also, with the influx of people after World War II and the advent of the air conditioners that made sprawling suburbs possible, civilization was moving in on a massive scale. No longer sleepy adobe towns lingering in the reveries of conquistadors and vaqueros, Phoenix and Tucson leapt their traditional bounds, spreading out in concentric rings of industrial-age glitter ever farther toward the sharp outlines of mountains. With that came factories and freeways, fast-food chains, swimming pools doused

The saguaro habitat, showing the yellow-blossomed brittlebush (foreground) and the red-blossomed hummingbird bush, or chuparosa *(background). Such desert plants thrive despite dry conditions.*

with chlorine, and new schools—all this apparently squeezed from the same molds used with abandon across the nation. Blessed by the hum of millions of whirling air conditioners, life on the desert became pretty much the same as it was elsewhere.

As far as the environment was concerned, the rapid growth marked a radical, a traumatic, disruption. Plainly, the desert was being pushed back, and with it the thrashers and phainopeplas, the Gila monsters and coyotes that once were backyard fare even for dwellers in the state's capital. Wells were failing, the very rivers drying up. As for the saguaros, in some areas, especially in the eastern portion of Saguaro National Monument, the young ones were not taking hold and the grand adults stood cracked and drooping, leaking a pungent, brown liquid. Elsewhere, people thought that a close relative, the organ pipe cactus, was also in trouble. Were they doomed? Perhaps in some sort of modern-day morality play, the moths and rust of civilization were creeping in with the overwhelming "progress" to destroy even the very symbols of the Southwest's deserts.

The idea had its morbid appeal: doom visited on a culture careless of its fragile natural heritage. And it made good headlines. Newspapers ran alarming articles: "Mystery Malady Attacks" and "Kings of the Desert Meet Death." Even scientists were puzzled. As for desert lover Krutch, he could only fumble, "I don't know why," then speculate, "I'm not sure if it's from the disastrously dry season we've had, or that they may be suffering from some of the smog that drifts down through that valley."

So the ball was thrown into the courts of the scientists, while an ecologically awakened public clamored not just for an explanation but for a cure. Yet science does not march to the nervous beat of headlines. And rightly so. It takes its time. It wants to gather all the facts, weigh and debate them, and debate them again, before making a pronouncement. Often in this painstaking process, it gropes up dead alleys, to return to the starting blocks and begin again. Responding to Krutch's suggestion that pollution was doing the damage to cacti, a scientist offered only this: "It has some effect, but we have no information that it is a real threat." Scientists may still be groping.

Standing skeleton of the giant saguaro

51

An old giant about to topple

In the case of the sickening saguaros, suggested directions for exploration, some of them deserving years of careful study, came thick and fast, both from professionals and laymen. It could be that civilization, intentionally or not, indeed was the villain in the continuing demise of the saguaro. After all, persistent as they are, saguaro seeds cannot prosper in the parking lots or golf courses now occupying their previous habitat. Maybe it is cactus pirates, who go about by night digging up all manner of what they see as spiny curiosities. Heinous as their crimes are — and as heavily penalized as an outraged public is demanding they be — such wantonness is little when compared to the daily — and entirely legal — obliteration of the landscape by the blades of bulldozers in the employ of freeway and subdivision builders. Whatever our pain at the loss, however, Arizona can still boast some rather heady expanses of wild, uninhabited desert. Even in some of these areas, far away from golf courses and pirates, the saguaros were, and are, in trouble.

The problem might lie with vandals, those individuals who try to compensate for a variety of impotencies through rifle and hatchet work. It is not uncommon to come upon a giant cactus, that emblematic hero, simply shot or hacked to pieces. But again, the damage for the most part is limited to roadsides, picnic, and camping grounds, areas accessible by car or jeep. For reasons not yet explained by psychology, hikers are of a healthier turn of mind.

Few backpackers lug rifle or axe for miles deep into the desert wilderness, enduring hours of thirst and pitiless sun, to simply stand alone in the middle of nowhere and blaze away at a saguaro.

While scientists lacked solutions, they faced no dearth of other suggested causes: woodcutting, fires, lightning, drought, a permanent change of the climate. All of them were investigated.

Back in the days before trains hauled in coal, and pipelines brought oil and natural gas to the state, almost all heating and cooking was done with wood. Hence, people needed a year-round supply. Especially valued was mesquite, a hot-burning, sweet-smelling fuel. Unfortunately, the species was one of the prime nurse trees sheltering the struggling saguaro seedlings from heat, cold, and the hooves of trampling cattle. Woodcutters denuded whole acres around settlements. Yet the destruction, while effectively eliminating saguaros, does not account for overall declining saguaro numbers elsewhere.

Neither does fire or lightning. After several good seasons of rain, the desert undergrowth flourishes. But in the dry season the new growth becomes a ready fuel, a pyre for the fire-vulnerable saguaros. As the tallest object in its natural surroundings, the cactus is vulnerable, too, to lightning strikes, which can have the effect on the saguaro of a handgrenade going off in a closet.

For all that, the saguaro has survived thousands of years of fire on the ground and from the sky.

This saguaro has been vandalized with a machete.

Similarly, it has made it with aplomb through droughts, shrinking as it draws on its internal water supplies but hanging on to swell back to normal shape when sufficient moisture arrives. Detailed meteorological records do not go back much more than a hundred years or so for Arizona. Perhaps, as some groups in California maintain, the earth indeed has shifted on its axis, bringing on extreme swings in the weather, which of late have sealed the giant's fate. The notion is doubtful, but if it is true, neither we nor the saguaro can do anything about it.

So scientists went down the list, eliminating the most visible explanations for saguaros giving up the ghost. One of the items loomed large both for its connection with the American psyche and its widespread and well-documented effects on the land.

Our view of the American West hangs on one peg: cattle. Without the horned, bawling creatures and the paraphernalia of ranches, cowboys, and winsome ballads, it is hard for us to imagine what the West would be like. Without the cattle that launched the romantic appeal of the open spaces, we would have no rodeos, no modern-day cowboy bars, no horse operas on late-night television—in short, no symbol of the six-gun-toting, free-as-the-wind exemplar of a hero so dear to the nation that Americans still cling to him with the desperation of people worshiping their childhood idols into adult-

Night lightning in the saguaro forest

hood. No matter that historians have shown the image false, a money-making creation of novelists and filmmakers. The big-hatted man on a horse keeps tugging at our hearts, and we slide over the fact that "western" writer Zane Grey was a dentist from Ohio who made a fortune by trading his drill for a pen dipped into romantic ink.

It is the effects of cattle on the western landscape that concern ecologists. The Spaniards brought the animals to the New World soon after their conquests began. As their empire expanded, they drove their herds north into what is now the American Southwest. Records of their impact is spotty, but an educated guess is that, given the sparse populations both in humans and livestock, cattle had fairly little impact on the Southwest under Spanish and Mexican rule, relatively speaking, of course. By the latter part of the nineteenth century, the United States had brought order to a land torn by Apache raids, making the wild territory safe for the extraction of silver and gold. But there were not enough precious metals to go around. What to do with this supposedly worthless desert but to turn beafsteaks-in-the-making loose on it?

So they came by the thousands, by the mooing tens of thousands, arriving by the trainload to wander at will, snuffing up the free grass from almost every nook and cranny of Arizona. Biologist David E. Brown comments, "By 1912, when Arizona and New Mexico attained statehood, the

Southwest had been thoroughly settled. Almost every live stream, every arable piece of flat ground, every meadow possessed a ranch or a farm. Use of the range was universal. . . ."

We have become so used to a desert paved by sand and stone that it is hard for us to picture that less than a century ago grass covered much of it. Grass that grew thickly above the ground and with a thick network of roots beneath. It was grass that held much of the arid Southwest together.

Suddenly stripped of this ancient cover, the land went into shock. Strong desert winds blew the thin but fertile soil away. Instead of soaking in, rains washed off, bearing the soil with them. Streams and rivers that once meandered through marshy areas boiled in brown floods during the storm seasons. Their churning waters ate at their banks. Areas that for centuries had been perennial watercourses began sinking into their beds. They then disappeared. In a land of little rain to begin with, the impact of cattle on the flora and fauna is incalculable.

The saguaro suffered along with the rest of the plant community. Cattle trampled seedlings and, through more trampling and browsing, destroyed the microenvironments under nurse trees for future generations of saguaros.

There is a more complex speculation, which remains a layman's conjecture only, that is worth mentioning as an example of one of the circulating theories attempting to explain the saguaro's trouble.

With the wrongheaded notion that coyotes fed on cattle, ranchers shot, poisoned, and trapped the little wolves at every opportunity. The truth is that coyotes feed largely on rabbits and other small fare. Rabbits in turn eat desert vegetation. The more rabbits, the more vegetation eaten. With the coyote population reduced, the rabbit population soared. So unwittingly the ranchers were robbing their own cattle of food while adding to erosion. But more to our purposes, coyotes also eat rodents. With the falling numbers of coyotes, rat and mice numbers increased, adding more clubs to the gauntlet that the tasty saguaro seedlings faced.

Still, as dramatic as all this sounds, it does not offer the explanation. Saguaros were failing for no apparent reason even where the government protected them — on reserves where there were few fires, no woodcutters, and, in cases, no cows.

Such a mystery puts the public on edge. Americans like to think that their technology can, like an everready genie, be called upon to straighten out whatever goes awry in their world. We resort to extreme measures the more the situation defies our efforts.

Back in the early 1940s, the visitor to Saguaro National Monument could have witnessed a shocking sight. Here on a reserve established to protect the tall cacti from the very ills mentioned above, men were attaching ropes to saguaros and pulling them over with trucks. They were not vandals or the notorious cactus pirates on the loose but government employees.

Destruction of a saguaro by jackrabbits seeking moisture stored in the cactus's tissue.

Once the giants thudded ignominiously to the ground, the men swarmed over them with axes and like samurai warriors gone berserk chopped their victims to pieces. Dump trucks next arrived to haul off the debris to a great pit. There the workmen carefully fumigated the jumbled remains with a mixture of paradichlorobenzene and kerosene. Lastly, they covered the grave with soil.

In subsequent years one might have witnessed an even more bizarre sight—a scientist armed with a great syringe climbing a ladder to give a massive injection of penicillin to a saguaro, or others assiduously spooning out pockets in saguaro trunks and dousing the holes with a solution of bleach and water.

The occasion of all this activity was a moth. Summarizing the situation, the superintendent of the Desert Botanical Garden in Phoenix concluded:

Research over the past decade shows that the saguaro is being somewhat depleted by a bacterium *(Erwinia carnegieana)* carried in the intestines of the larvae of a moth *(Cactobrosis fernaldialis)*. This moth lays her eggs in the tissue of the saguaro and as the larvae eats the soft tissue of the saguaro the bacterium is then spread through these tissues and thus spread through the entire plant. This can be a slow process in the summer months but during the rainy period the bacterium spreads very rapidly.

The superintendent did not remark on the irony that a microorganism named for Andrew Carnegie was also attacking the more famous namesake of the philanthropic plant lover, but he did go on to observe that the disease usually sunk its teeth into older cacti, those less vigorous and hence already "weakened due to age," a point to keep in mind.

So scientists had at last, through diligent research and a meticulous process of elimination, found the Achilles' heel of the sick and dying giant cactus. They had finally identified the vulnerable spot threatening to bring down a plant empire. The solution appeared simple: tear down the thoroughly infected plants and disinfect the afflicted parts of others, or stop the disease in its tracks with penicillin. Not easy to do over thousands of square miles studded with millions of spiny giants, but possible, if we truly wanted to save the Southwest's best-known emblem.

Nonsense, said other researchers. For if scientists insist on making minute inspection of all the evidence before making their pronouncements, sometimes, even after that long process, their pronouncements are at loggerheads. Since they first entered the Southwest, English speakers have noted saguaros in chopfallen states of wilt. Yet if not always flourishing over some of their territory, they have prospered in other parts. What really afflicts the saguaro has little to do with beef on the hoof, or woodrats, or a rot-producing bacterium with the jawbreaking label of *Erwinia carnegieana*. In fact, what has to do with the drooping giants, according to plant ecologists, is entirely beyond the control of humans. Simply put, occasional cold spells limit saguaro numbers, periodically nipping at the cactus in the coldest parts of its range.

For an explanation, we need to look briefly at where saguaros grow and, in particular, at the places where they grow the best. The Sonoran Desert covers much of south-central and western Arizona and the western portion of Sonora, Mexico. Here the saguaro is coterminus with this major arid region. However, a corner of this desert laps over into southern California, then follows most of Baja California, almost to the peninsula's tip. In this second portion, with a few exceptions, the saguaro is absent.

For more than a century, scientists have known the factors that regulate a plant's range. Each species depends upon a certain combination of correct soil, moisture, exposure to sun, etc., in order to survive. These are the limiting factors determined by the plant's genetic makeup. Because of this, ironwood trees cannot grow in Massachusetts due to the cold any more than elm trees could endure the desert's summer heat. As far as the saguaro is concerned, its range extends across the Colorado River for a little ways into California, but the plant appears only in small numbers and in just a few locations, for that part of the Sonoran Desert receives too little summer rain to support even the water-saving giant.

As mentioned earlier, the visitor may note that saguaros grow on one side of a hill but not on the

Some saguaros grow in an atypical habitat of sand, such as this saguaro forest near the Pinacate Mountains of northern Mexico. Note the cristate saguaro left of center.

other. It should also be noted that even within its wide range, the cactus marches up the sides of mountains only to stop abruptly and let the mountains go on without them. One might even chance on a grove of saguaros backed up against a cliff face doing just fine, while none grow out in the open.

In all these cases, water is not the limiting factor. A little thought on the last example brings the answer. The rock of the cliff absorbs the sun's heat, heat slowly released to the cacti throughout the desert's cold winter nights. Similarly, the south slope of a hill receives and retains more sunshine than the north side, and, generally, the higher the elevation, the cooler it gets. As far as the saguaro is concerned, cold is the chief key in limiting its northern and eastern range in Arizona.

For when hard freezes hit the Sonoran Desert, as they do now and again in these portions, the cold damages saguaros and may leave them with drooping arms, and if severe enough, it kills them outright. In this case, rot sets in, the cacti sicken and begin leaking a tell-tale putrid liquid.

The hitch in the scenario is that saguaros sometimes give up the ghost slowly. They can be snowed or sleeted on, frozen in layers of ice, but can take up to a decade to keel over from the injury. By then, the wondering observer sees no connection between the dead and bacterium-ridden mass on the ground and the cold snap of years earlier. But if records are checked, one will find extreme dips in temerature occurring periodically, for instance in 1870, 1913, 1937, 1962, 1971, and 1974, establishing a correlation in saguaro deaths. Die-offs, then, are normal. The saguaro struggles to hold its own in some parts of Arizona not because of the outbreak of a disease but because those sections are colder than others. The bacterium establishes itself after a freeze has done its damage. Therefore, bacterial activity is a result of, but does not cause, the death of saguaros.

In short, the saguaro is a subtropical plant, in the coldest parts of its range subject to periodic attacks by cold and hence limited geographically by temperature lows. The tall cactus has lived with cold spells for centuries, probing into new territory during warm times, only to be driven back by freezes. Like us, and indeed like all plants and animals on the planet, the saguaro is still evolving, still adjusting to ever changing circumstances. Eventually, it may extend its range, or it may stay pretty much where it is, clinging tenuously to the colder fringes of its habitat. Only the centuries will tell.

The one thing that the saguaro cannot adjust to, however, is human abuse. Today, the saguaro still marches over the lava hills toward endless horizons. Such scenes offer visual spectacles for visitors. More subtly, the tall cactus serves as a complex ecological

Snow in the saguaro forest

focus in the lives of insects, lizards, woodpeckers, coyotes, and many other creatures of the Sonoran Desert. But those breathtaking sweeps and all they contain are not endless, not limitless, as our ancestors once thought. Today in the booming cities of the desert, the growing human population gnaws away, sometimes with madcap speed, at the saguaro forest, even while farsighted citizens urge preserving as much as possible of this unique forest in its natural state. Only our children will know how well we have cared for the desert heritage that we pass on to them.

Further Reading

As indicated in the foregoing, knowledge about the saguaro continues to expand, yet as it does, experts also continue to disagree about some aspects of the giant cactus. However, readers who wish to explore the saguaro further on their own will find the books below in the main authoritative.

Ruth Kirk offers a good general introduction to the arid parts of the Southwest in *Desert* (Boston: Houghton Mifflin, 1973). Her information on the bacterium that attacks saguaros, however, is dated. On this score the reader might turn to the concise article on the tall cactus in Lyman Benson's fine guide, *The Cacti of the United States and Canada* (Stanford, California: Stanford University Press, 1982).

Two other serviceable overviews of the continent's deserts are Edmund C. Jaeger's *The North American Deserts* (Stanford, California: Stanford University Press, 1957) and Peggy Larson's *Deserts of America* (Englewood Cliffs, New Jersey: Prentice Hall, 1970).

As for Indians and the saguaro, Gary Nabhan offers delightful going for the casual reader interested in a scientist's impressions of the warm Papago people and their bond with the earth in *The Desert Smells Like Rain: A Naturalist in Papago Indian Country* (San Francisco: North Point Press, 1982). Nabhan takes readers to Papago sacred sites, comments generally on the changing culture, and participates, somewhat hilariously, in the traditional wine-drinking ceremony.

Two long anthropological studies, including sections on Papago myths, dances, and songs, reveal the richness and complexity of traditional Papago life, much of it revolving around the saguaro. They are Frank Russell's *The Pima Indians* (Tucson: University of Arizona Press, 1975) and Ruth M. Underhill's *Papago Indian Religion* (New York: AMS Press, 1969).

Nonspecialists, however, might first turn to Frank S. Crosswhite's *The Annual Saguaro Harvest and Crop Cycle of the Papago, with Reference to Ecology and Symbolism* (Superior, Arizona: Boyce Thompson Arboretum, 1980. Published as a special issue of *Desert Plants* 2, Spring 1980). Despite its rather forbidding title, this offers a readable survey of the saguaro in Papago life, with emphasis on the saguaro harvest and the following ceremonies. Bernard L. Fontana's concluding bibliography points toward further sources. One of these deserves mention. L. S. M. Curtin expands on the catalog of plant uses among the Papagos and their relatives the Pimas in *By the Prophet of the Earth: Ethnobotany of the Pima* (Tucson: University of Arizona Press, 1984).

For the technically inclined, Charles H. Lowe and Warren F. Steenbergh detail years of research in *Ecology of the Saguaro*, a three-part presentation concerning climatic conditions, reproduction, survival, and demography relating to the giant cactus ("Ecology of the Saguaro: I." In *Research in the Parks.* Washington, D. C.: U. S. Department of the Interior, 1976. National Park Service Symposium Series, no. 1; *Ecology of the Saguaro: II.* Washington, D. C.: National Park Service, 1977. National Park Service Scientific Monograph Series, no. 8; and *Ecology of the Saguaro: III.* Washington, D. C.: National Park Service, 1983. National Park Service Scientific Monograph, no. 17).

Yet perhaps the best thing to do is to stroll out among the giants, into the saguaro forest itself. Listen to the wind humming through thousands of spines. And far from civilization, watch for a flicker bringing food to her young high in a saguaro trunk, while as a background, great white ink spots of clouds drift up from Mexico.

Peter Wild came West at age eighteen to become a cowboy, but quickly switched from that to English studies. He is currently teaching creative writing — prose and poetry — at the University of Arizona in Tucson, where he has been a full professor since 1979. He is the author of over thirty poetry books, ten nonfiction books, one hundred articles, and two hundred book reviews appearing in such periodicals as *American West, Backpacker, Sierra Club Bulletin,* and *Smithsonian.*

Hal Coss is retired from the National Park Service, for whom he worked as a park ranger, chief park naturalist, and resources management specialist. Most of his career was spent at Organ Pipe Cactus and Saguaro national monuments. His photographs have appeared in numerous publications. He is working temporarily as a research analyst in the School of Renewable Natural Resources, College of Agriculture, University of Arizona.

WESTERN HORIZONS

Vignettes of the western experience

Also in this series

IN THE PATH OF THE GRIZZLY

Text and photographs by Alan Carey

A photographic record of the grizzly is combined with the author's observations on the bears' behavior for an insightful glimpse into grizzly life.

84 pages · 35 color photos · ISBN 0-87358-394-9 · sc · $11.95

MINIATURE FLOWERS: A Desert Search

Text and photographs by Robert I. Gilbreath

These delicate flowers measure approximately four millimeters in diameter and have unusual adaptations that enable them to survive harsh desert conditions.

84 pages · 35 color photos · ISBN 0-87358-382-5 · sc · $9.95

THE TRUMPETER SWAN

A White Perfection

Text and photographs by Skylar Hansen

Four nesting pairs of swans are followed throughout their seasonal activities of nesting, raising the cygnets, and winter survival.

84 pages · 45 color photos · ISBN 0-87358-357-4 · sc · $9.95
ISBN 0-87358-358-2 · hc · $14.95

WILDERNESS ABOVE THE SOUND

The Story of Mount Rainier National Park
by Arthur D. Martinson
Foreword by Alfred Runte

This photodocumentary of the history of Mount Rainier and surrounding environs includes a stunning collection of historical photographs.

96 pages · 7 color photos · 40 halftones
ISBN 0-87358-398-1 · sc · $11.95